THE DOMINIE WORLD OF
OCEAN LIFE
ANTARCTICA

Antarctica
The Ice Continent

WRITTEN & PHOTOGRAPHED
BY KIM WESTERSKOV, Ph.D.

1. Ice, Snow, and Sky 2
2. Icebergs 6
3. The Birth of an Iceberg 8
4. Glacier Tongues and Ice Shelves 10
5. Sea Ice 13
6. Fast Ice 15
7. Camped on a Frozen Sea 17
8. Animals of the Sea Ice 20
9. Under the Sea Ice 24
10. Icebreaker to Antarctica 27
11. People, Ice, and Icebergs 29
    Glossary 32
    Index .... Inside back cover

Dominie Press, Inc.

1

Ice, Snow, and Sky

When we reach the Antarctic continent, we find that most of the "land" is not rock or soil or sand—but ice. Nearly everything we look at is ice. If we walk or fly inland, we often see nothing but ice. In many areas, you can fly 1,000 miles or more in one direction and see only ice, snow, and sky—white and blue.

Occasionally a peak or mountain range pierces the ice, but 99 percent of **Antarctica** is covered by ice. This permanent ice sheet contains 90 percent of the world's ice and 70 percent of our planet's freshwater. Antarctica contains about 7 million cubic miles of ice.

Ice dominates the seas, too. Surrounding Antarctica is a huge ring of sea ice, which is floating frozen seawater. At its maximum in late winter, the **pack ice** alone covers an area twice the size of the entire United States, in a

▲ *Antarctica: a portrait in white and blue.*

layer mostly no more than three feet thick. At its narrowest, this belt is about 350 to 400 miles wide; at its widest, it is 1,800 miles across. The Antarctic continent itself measures 5.4 million square miles, about the size of the United States and Mexico, combined.

Like floating castles, **icebergs** drift with the pack ice. There are hundreds of thousands of them, some as large as Jamaica, but

NOTES

As our ship travels south toward Antarctica, we watch albatrosses soaring over restless seas and feel the sharp wind growing colder and colder. The first real sign of Antarctica is a lone iceberg, shining white in the sunlight. Soon there will be ice in many shapes and sizes. Like snowflakes, no two icebergs are exactly alike.

most are less than half a mile long. The most dangerous icebergs are even smaller. Growlers, car-sized icebergs, and bergy bits, icebergs the size of a small house, are too small to be picked up by radar or by a ship's watch. Unfortunately, they can rip open a normal ship's hull. Most of an iceberg is below water—exactly how much depends on the density of both the portion above water and the portion below water.

Although they are often beautiful and dramatic, icebergs are dying floating islands of ice. ▶

Even small icebergs like this one pose a danger to passing ships. ▼

2

Icebergs
Beautiful and Dangerous

Icebergs are large floating pieces of freshwater ice that have calved, or broken off, from **glaciers**, ice sheets, or **ice shelves**. Between 10,000 and 50,000 icebergs are created in the **Arctic** each year, most of them from the glaciers of Greenland. In 1912, a Greenland iceberg drifted south into "Iceberg Alley," east of Labrador, and collided with the *Titanic*. The "unsinkable" ship sank within three hours with the loss of over 1,500 lives.

◀ *A large tabular iceberg floats behind a small iceberg.*

Antarctica produces many more icebergs than the Arctic, and far bigger ones. We can only guess at the number, but they add up to about 350 cubic miles of ice in most years. That's 90 to

The fast ice surrounding this iceberg is breaking out due to the combined effects of currents, tides, winds, and "warm" summer temperatures. ▶

95 percent of the world's total volume of icebergs each year. The ice cliffs around most of the Antarctic coastline are the seaward edge of huge ice shelves. These often break off in much larger chunks than the icebergs from the much smaller valley glaciers.

The huge icebergs from ice shelves are called tabular, or table top, icebergs. Like the ice shelf from which they broke off, they are thick and flat-topped. One of the largest icebergs ever seen broke off Antarctica's Ross Ice Shelf in March 2003. It was 185 miles long and 23 miles wide. It had an area of about 4,250 square miles, a little bigger than Hawaii. Storms and currents have since broken the mega-berg into smaller pieces.

How long do icebergs live? Most melt within a few years, but some live for up to twenty years or longer. Of the hundreds of thousands of icebergs that calved the year you were born, most would have broken up and melted by the time you started school.

3

The Birth of an Iceberg

Antarctica and the Arctic are very different from each other. The Arctic is an ocean around the North Pole surrounded by land. The ocean is covered by ice all year, and snow and ice lie on the land for most of the year. Antarctica is land around the South Pole surrounded by ocean.

Covering Antarctica and Greenland, our planet's two main sources of icebergs, are huge ice sheets containing most of the Earth's freshwater.

The thickness of this huge ice sheet is hard to imagine. It is up to three miles deep in places—about half the height of Mount Everest. The thickness of the Antarctic ice sheet makes Antarctica the highest continent. This is the main reason why it is much colder than the Arctic. The

◀ *The layers in this grounded iceberg tell the story of how snowfall after snowfall has slowly been compressed into ice. Each layer often—but not always—represents one year's snowfall.*

▲ *As ice moves downhill under its own weight, it moves around or over any obstacles. Like river rapids in slow motion, pressure within the ice breaks it up into heavily crevassed areas.*

Glaciers are frozen rivers that carry ice downward. Each one is different. ▶

higher you go above sea level, the colder it becomes.

This ice sheet has been built up over millions of years by snowfalls and ice crystals that fall out of the clear blue sky. Most people think of Antarctica as a place where it snows a lot, but that's only true around the coastlines. The high polar plateau is actually a desert.

The largest glacier on Earth is the Lambert Glacier in Antarctica. That glacier is 250 miles long and 30 miles wide.

4

Glacier Tongues and Ice Shelves

If a helicopter dropped you for a few hours somewhere in the interior of Antarctica, the main things you would notice would be how cold it was, how empty it was, and how quiet it was. In most places all you would see would be snow, ice, and sky. Apart from the wind, nothing would seem to be moving. If you stood there on the ice long enough, you might hear occasional creaks or groans from deep inside the ice, clues that perhaps something was moving. In fact, most Antarctic ice is moving, although slowly. If you planted a flag exactly at the South Pole, in one year's time it would move about thirty feet.

Ice streams move much faster than the rest of the ice through which they move.

NOTES

Here I am deep inside an ice cave in a floating glacier tongue. It is spring and still very cold, so we can safely travel on the fast ice. It is both very quiet and very beautiful in here. Huge delicate ice crystals grow from the ceiling in the dry, still air.

▲ *Ice shelves are usually white because they are covered by snow, but the part of the McMurdo Ice Shelf called the Dirty Ice is far from white. Summer ponds and patches of dirt, diatoms, bacterial mats and seafloor animals make this a very unusual and complex area. It is midsummer, and all the sea ice has disappeared.*

It is hard to tell where glaciers and ice sheets let go of the seafloor and start floating as glacier tongues and ice shelves. ▶

And ice streams are huge—up to 50 miles wide and almost 500 miles long. These ice streams drain the ice from the high Antarctic ice sheet and move it downhill toward the coast. There, it joins up with ice that has arrived by glaciers or other ice streams to form huge, wide sheets of very thick ice.

5

Sea Ice

There are two main kinds of sea ice: **fast ice** and pack ice. Fast ice is the unbroken ice that is firmly attached to the shoreline or ice shelves. Outside this lies the pack ice, a changing, moving mixture of broken ice and open water.

In late summer, Antarctica's sea ice is at its minimum size, with an area of about 1.5 million square miles. As autumn and winter tighten their icy grip, the open sea freezes over at the amazing rate of about 25 square miles every minute, or up to 40,000 square miles every day. Later, in spring, it will melt twice as fast.

This is an aerial view of new sea ice forming beyond the edge of last year's fast ice. ▶

◀ *The pack ice with open water beyond it. Ice floes, flat sheets of sea ice broken into pieces by wind and currents, come in all sizes.*

By the end of the Antarctic winter, in September, the sea ice reaches its maximum size, about 8 million square miles. Like the cold breath of a vast ice animal breathing in and out just once a year, the Antarctic pack ice appears and disappears every year. Some scientists call it the greatest seasonal event on Earth.

The winter pack ice cuts Antarctica off from the rest of the world. No ship, not even a large modern **icebreaker**, can make its way through to the coastline. But even in the coldest winter, Antarctica is never completely surrounded by solid ice. Winds, ocean currents, and the shape of the coastline and sea floor combine to keep the ice on the move, breaking it up into **ice floes**, and creating open-water areas.

The thickness of the Antarctic sea ice depends on where it is, and how long it has been forming. The ever-moving pack ice thickens only to about three feet. This is the result of wave action and ocean currents, and because it rarely lasts more than one year. The fast ice clinging to the coastline thickens to about eight feet, or even thicker if it does not melt during summer.

6

Fast Ice

It is early summer and a US Air Force Galaxy C-5—the world's second-largest aircraft—is landing on the McMurdo Sound fast ice. The nearest solid land is the seafloor, 1,300 feet below, but the Galaxy is perfectly safe on this sea ice runway. When a 300-ton, fully loaded Galaxy is sitting on the thick fast ice, its wheels depress the ice only three inches. That's how strong the fast ice is.

A US Air Force Galaxy C-5 aircraft on the sea ice runway in McMurdo Sound. ▼

The fast ice near the Antarctic coastline is often flat, hard, and smooth—ideal for spring and early summer travel by snowmobile or larger vehicles—even for large aircraft to land. The main dangers are areas of rough ice, thin ice, and huge cracks. ▶

▲ *Dramatic sea ice pressure ridges are often taller than a person. They sometimes look like frozen waves, which is more or less what they are—lines of jumbled sea ice pushed against the shoreline.*

Fast ice is sea ice that forms a sheet attached to land or ice shelves. In some areas it can be thirty miles wide, but usually it is much narrower. Most of it is flat and smooth, and it often has a thick covering of snow. Rough ice, or broken ice, is the result of floes freezing together during the winter freeze, or when the ice is piled up in jumbled heaps by waves or ice pressure.

Fast ice develops around the Antarctic coastline during winter, especially in bays and shallow areas. Most of the fast ice has broken up and moved away by late summer, except in some bays. Antarctic fast ice normally reaches a thickness of six to eight feet—twice the thickness of the pack ice—but in some areas, multi-year ice may build up to a thickness of up to twenty feet.

7

Camped on a Frozen Sea

Wind is a killer in Antarctica. With good clothing, working outside at temperatures of minus thirty degrees Fahrenheit, or even minus sixty, presents no great problems. But as soon as the wind springs up, these same temperatures become dangerous. When **blizzards** rage, humans simply cannot survive without shelter, no matter what they are wearing.

The McMurdo Sound fast ice is the skin of frozen sea joining Ross Island to the continent of Antarctica. During spring and early summer, one of the best ways of getting around the Antarctic coastline is by traveling on the fast ice. The fast

NOTES

We had planned to sleep in a hut on the Antarctic mainland that night, but our all-terrain vehicle had broken down so we pitched camp on the sea ice. Just six feet below our polar tents was deep, dark, very cold seawater, but we were perfectly safe. Wearing most of my clothes and tucked inside a double sleeping bag, one sleeping bag inside another sleeping bag, I was cozy and warm.

NOTES

I will never forget that evening meal on the sea ice. It was spaghetti and mince, lovely and hot when it hit my plate. By the time I was half way through, my dinner was stone cold. And by the time I reached the other side of the plate, the food was frozen solid. I couldn't even scrape it off my plate. The temperature was minus twenty degrees Fahrenheit. At times like those I thought about the early Antarctic explorers. They were here for years at a time, out in all weathers, and without weather forecasts, radio, or warm, modern clothing. Life for them must have been awful at times.

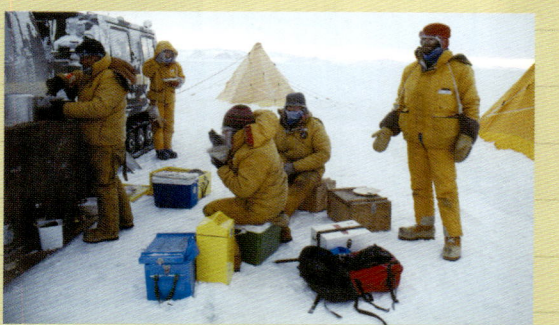

ice is mostly flat, smooth, hard, and very strong, and so it makes an ideal runway for aircraft. In summer, scientific parties traveling on the sea ice would need to know how thick the ice was this year, and what condition it was in. A team of researchers drilled many holes through the ice to measure its thickness, and mapped areas of safe travel, areas of rough ice, and all the dangerous cracks.

Shaped by sun and wind, this piece of broken sea ice rests on the sea's frozen surface in McMurdo Sound, near the Antarctic coastline. ▶

Antarctic weather is a harsh reality for visitors, even when they are covered by protective gear. ▶

8

Animals of the Sea Ice

During the southern summer, Antarctic seas are some of the richest in the world, and countless whales, seals, **penguins**, and other birds come to feast on the plentiful food, especially the vast swarms of **krill**. These krill feed the largest animal ever to have lived on our planet, the blue whale. Most of these animals are summer visitors only, and by autumn they are gone. Antarctic winters are extremely harsh. It is dark twenty-four hours a day, and blizzards rage through the freezing blackness. Temperatures around the coastlines drop as low as minus seventy-six degrees Fahrenheit.

    The list of winter residents is short: some tiny insects, some small plants, **Weddell seals**, and emperor penguins. Nothing else lives there, apart from humans in their heated buildings. When the winter blizzards

▲ Adèlie penguins do a lot of walking on sea ice. Heading toward their colonies in early spring, they often need to cross up to sixty miles of fast ice. They walk, hop, or toboggan on their stomachs, averaging two miles per hour.

This is the emperor penguin colony on the fast ice at Cape Crozier, Ross Island. This is the world's southernmost emperor penguin colony. ▶

9
Under the Sea Ice

Diving under the Antarctic sea ice we find ourselves in a cold, strange world, but one with a wild beauty all its own. If we dive under the fast ice in spring and early summer, before the **plankton** blooms, we dive in what may be the clearest water in the world. It can also be very dark if there is a lot of snow lying on the sea ice. This makes it a little spooky, too, especially if you are diving in very deep water and there is only one way back out—the hole in the fast ice that you came in through.

As with most food chains in the sea, the Antarctic ecosystem starts with tiny drifting plants called **phytoplankton**, "drifting plants."

NOTES

My best memories are of the clear water, the beauty of ice, and swimming with Weddell seals. These animals might look clumsy and overweight when they're lazing on the sea ice, but underwater they are graceful and totally comfortable in the freezing water. Since they have no natural predators, Weddell seals are easy to approach.

24

▲ *By late summer, there is often no fast ice around the Adèlie penguin colony, though there is sometimes a jumble of rough ice pushed against the shore.*

◀ *Weddell seals, like this mother and her pup, depend on breathing holes and spend much of their time keeping them open. In winter, these holes are their only way of getting between the sea, where they shelter during blizzards, and the air.*

rage, Weddell seals stay underwater, coming up only to breathe. Up on the sea ice, emperor penguins huddle together in sleepy groups.

Emperor penguins are amazing birds that have successfully adapted to the toughest weather on Earth. All except two of the forty or so known **colonies** are not on land but on the fast ice, so most emperors probably never step onto real land during their entire lives. The sea ice plays a huge part in their lives. They breed and rest standing on the sea ice, but all their food comes from the sea: fish, squid, and krill.

The only other penguin that lives right around Antarctica is the Adèlie penguin. This much smaller species breeds on rocky coastlines, but spends much of its time walking on sea ice, shuttling between the colony and the open sea. It spends the winter in the pack ice north of the continent.

Most of the ten species of whales found in Antarctic seas are summer visitors only. In autumn they migrate north to their breeding grounds in **tropical seas**.

▲ Sunlight shines through seven feet of sea ice, giving it a beautiful blue color. The delicate ice is called platelet ice.

◄ A Weddell seal pup and its mother are in a long, wide crack in the sea ice. The surface water is filled with slush ice and small chunks of floating ice.

This is what the underwater side of a glacier tongue or an iceberg looks like.

Nearly all phytoplankton in Antarctic waters are diatoms, tiny plants living in boxes of silica. In the fast ice, diatoms live not in the seawater but in mats on the underside of the sea ice. In the pack ice, diatoms live inside the ice, in concentrations among the highest ever recorded in any ocean anywhere.

Feeding on these tiny plants are small animal plankton, or zooplankton, and these are in turn eaten by larger animal plankton. The main animal plankton in Antarctic seas is Antarctic krill, a shrimp-like animal about two inches long. Krill form dense swarms, with up to a million in every cubic yard. These swarms can be huge, up to 170 square miles in area. The Antarctic krill is the main food of squid, fish, penguins, seabirds, seals, and whales.

10

Icebreaker to Antarctica

A few days before Christmas, I joined the U.S. Coast Guard icebreaker *Polar Sea* in Australia. It would be my third visit to Antarctica that season, where I was taking 24,000 photos for the Antarctic Visitor Center in Christchurch, New Zealand. From the icebreaker, I hoped to photograph albatrosses, penguins, sea ice, icebergs, Antarctica from a helicopter—and maybe a real Antarctic storm. We would be sailing through the roughest ocean in the world, the Southern Ocean.

As we approached Antarctica we entered the pack ice, with dense fields of ice floes as far as I could see. We steamed through

Breaking a channel through the fast ice: The reinforced hull of the icebreaker is designed to ride up on the ice, which is then broken by the ship's weight. The ship supports scientific research, carrying two helicopters and providing laboratories and accommodations for up to twenty scientists. ▶

this flat, icebound sea for a day. Penguins rested on the ice floes between dives. After awhile I began to notice that the pack ice was not flat anymore—it was gently rising and falling as ocean swells passed through it. Just a little at first, then higher and higher, until the ice floes were rising and falling fifteen feet. Suddenly the *Polar Sea* burst out of the pack ice into open sea—and a fierce storm.

    We were now in the Ross Sea, the world's southernmost open sea, about half the area of Europe. Each time our ship crashed into a swell, a wall of gray seawater rose up into the air and then hurtled toward us on the bridge. We all ducked. On an earlier trip, a huge wave broke the thick reinforced glass windows on the bridge. Life was not very comfortable onboard the *Polar Sea*, but we were safe, dry, and warm. Antarctic petrels soared over the ice-choked wave crests. I wondered how the penguins out there were doing.

NOTES

I quickly settled into shipboard life, living with 230 other people in a ship 399 feet long. Christmas Day was spent with my new shipmates. Both the air and sea were getting colder every day. Albatrosses and smaller seabirds soared and skimmed over a restless gray ocean.

11

People, Ice, and Icebergs

In Antarctica, ice makes the rules. It decides where we can build our bases, how and when we get to them, and how we get around. Moving people and supplies by aircraft is expensive, and often either dangerous or simply impossible. The only other way is by ship, and ships can only reach Antarctica for a few brief months each summer.

Antarctica is a long way from where most of us live, but it still affects us in many ways. The more we learn about Antarctica's ice and its weather systems, the more we realize

A cargo vessel and a fuel tanker forge through fast ice opened by an icebreaker. ▶

◂ McMurdo Station. A U.S. container ship unloads its cargo onto the floating ice wharf, which was made by artificially thickening the sea ice the previous spring and early summer.

that Antarctica affects the rest of the world—especially its weather.

It works the other way as well: what happens in the rest of the world affects Antarctica. If global warming results in winters with less sea ice than normal, then there are far fewer krill. One or two seasons may not make a big difference, but several winters with poor ice will result in a big drop in the krill population. Fewer krill affects most Antarctic animals, which rely on krill as a source of food. For example, less food for penguins means fewer chicks will be raised.

NOTES
Antarctica doesn't belong to any nation or nations; it's ruled cooperatively by the nations that signed the Antarctic Treaty in 1959. It could become a place where we repeat the mistakes we've made in the rest of the world—or it could remain a continent of wild beauty, a continent of peace and cooperation. As with so many things, the choice is ours.

▲ *This is a sight you won't see anymore. By international agreement, the last huskies were taken out of Antarctica in 1994.*

Glossary

Antarctica: An uninhabited continent surrounding the South Pole

Arctic: The cold, barren region surrounding the North Pole

Blizzards: Severe snowstorms with powerful winds

Colonies: Groups of animals of the same kind that live together and are dependent on one another

Diatoms: Single-celled algae with silica-filled cell walls

Fast Ice: A patch of unbroken sea ice attached to the shore

Glaciers: Large bodies of ice and snow that originate on land and slowly move toward the sea under the pressure of their own weight

Ice Floes: Large, thick sheets of floating ice

Ice Shelves: Thick masses of ice covering a coastal area and extending into the sea

Icebergs: Huge mounds of ice that have broken away from a glacier and float in the open sea

Icebreaker: A ship with a reinforced bow used to cut through thick ice and create a passage through frozen seas

Krill: A very small shrimp-like marine animal that is a primary source of food for penguins, whales, and many other inhabitants of Antarctica

Pack Ice: An area of broken ice and open water

Penguins: Flightless, web-footed seabirds that use their flipper-shaped wings for swimming

Phytoplankton: Tiny free-floating algae that live near the surface of seas worldwide

Plankton: The small animal and plant life in a body of water

Tropical Seas: Bodies of water that are very warm throughout the year

Weddell Seals: Marine mammals that live under the ice surrounding the Antarctic coast